servicenw

certified system administrator
study guide

Authors' Profile

Muhammad Zeeshan Ali
PMP, PMI-ACP, CSA, OCP

Enterprise Agile Coach | Trainer | Organizational Change Builder | Author | Certified Agile Practitioner

Saqib Javed John
PMP, PMI-ACP, CSA, ITIL, SCJP, SCWCD, FIDIC

Author | Trainer | Certified Agile Practitioner | PMO | Organizational Development Consultant

Author of multiple books and numerous articles elaborating new dimensions of Agile framework and Traditional Project Management along with his work on Performance Management, PMO, Leadership, Team Building and Personal Motivation. He is best known for designing the first of its kind "Performance Measurement Matrix" to calculate number-based performance indicators and scoring for both Software Engineering Individuals and Teams. Zeeshan is a great advocate and promoter of adaptation of Agile Methodologies, Processes and Team Skill building.

Zeeshan has over 20 years' experience of managing 100+ mid-large scale, high visibility projects in both the Public and Private sectors. Experienced in managing several significant projects simultaneously and with a team spread over different geo-locations.

Zeeshan has Degrees in Project Management (MS) and Computer Sciences (BS). He has been certified as a Project Management Professional (PMP) and Agile Certified Professional (PMI-ACP) by Project Management Institute (PMI), USA.

Saqib is one of the founding members and Managing Director of Organizational Governance Management Consultants (OGMC). He has professional expertise of more than 18 years of working on enterprise projects in various business domains ranging from a functional organization to projectized organization.

Saqib has immense experience in developing and managing human behavior, process engineering and optimization, risk management, conflict management, performance maturity audits and policy-making. This is one of the reasons he is relatable to readers of Business and management professions. He is best known for his rapid-learning techniques and easy methods of practical implementations. He also has contributed to many anthologies. His work is helping thousands of students, teachers and professionals.

Saqib is MS (IT), certified "Project Management Professional" (PMP) and "Agile Certified Practitioner" (ACP) from Project Management Institute (PMI) USA. He is also certified in "Information Technology Infrastructure Library" (ITIL) from Exin UK, "Sun Certified Java Programmer" (SCJP) and "Sun Certified Web Component Developer" (SCWCD) from Sun Microsystems USA.

Zeeshan and Saqib are also the Authors of **Exam Prep Series of PMP, ACP, RMP, ITIL** and other Management related books like **Inside Familiar Management, Private Life of Management, Applied Psychosomatic Planning, Agile Beyond Boundaries** to name a few, which are gathering a lot of attention among the students, practitioners and professionals of business and management sciences. All their publications can be reviewed at ogmcpublications.com

Both actively publish their exclusive articles and their blog can be reached at blog.ogmcs.com

servicen○w
certified system administrator
study guide

Muhammad Zeeshan Ali
PMP, PMI-ACP, CSA, OCP

Saqib Javed John
PMP, PMI-ACP, CSA, ITIL, SCJP, SCWCD, FIDIC

2022

All inquiries should be addressed to (e-mail): info@ogmcpublications.com

First Printing: 2022

ISBN: 9798370055607

OGMC Publications
ogmcpublications.com

Ordering Information:
Special discounts are available on quantity purchases by corporations, associations, educators, and others. For details, contact the publisher at the above-listed address.

Dedicated to all the readers

And

those who inspired this work.

Preface

This handbook is extremely different from any other book in the market, stores and libraries. In fact, a unique of its kind, giving a change to readers to quickly learn about the key concepts of ServiceNow.

This book is the shortest way for the students and professionals who wants to enhance their knowledge, skills and expertise on the ServiceNow Application, its Framework and Implementation. This book not only aims to provide quick career boost in a lesser time but also a smartest master piece to the ones who wants to appear in "Certified System Administrator" (CSA) exam in near future or later.

The notes are deliberately designed in form of the points to keep the readers interest alive. 98% points are directly related to the CSA exam but there are few which added to cover relevant knowledge.

ServiceNow is one of the dominant and globally known applications for managing IT Services Management, IT Operations Management, IT Business Management, IT Assets Management, GRC (Governance Risk and Compliance), SecOps (IT Security Operations), CSM

(Customer Services Management), HRSD (HR Service Delivery) and much more.

ServiceNow has very interactive, scalable, revolutionary and dynamic approach for managing IT infrastructure, operations, process flows and overall Information management.

The authors of this book have provided all the information based on their personal knowledge and experience with good intentions. However, we strongly recommend consulting other books and information material to maximize your chances of success in the CSA or any other ServiceNow exam.

Table of Contents

Why ServiceNow is a Good Career Move (Top 10 Reasons)

1. It gives you more functionality, dependability, and personalization options, also a well-known Enterprise Ticketing Tool and a significant participant in the ITSM space

2. ServiceNow offers technical management support to large IT operations, with a focus on the help desk and ITSM

3. ServiceNow has created a new line of work for IT professionals and has ability to grow and expand is a desirable feature

4. ServiceNow is a cloud-based company that dominates the IT service management industry and helps in managing and maintaining IT infrastructure — a key role for today's leading enterprises

5. It has a stronghold in the market and generates a lot of revenue because of its effectiveness

6. IT firms can collaborate with various departments on a single platform with ServiceNow, saving time and money

7. ServiceNow is an essential platform for cloud-based operations and consequently one of the most crucial business strengths an organization can possess

8. ServiceNow has been used by companies in nearly every industry that provides service and support, such as the automobile industry, IT industry, medicine, insurance, entertainment and media, and so on

9. ServiceNow and its cross-departmental capabilities, unique workflow, and data and process automation have made it a must-have for all enterprises

10. ServiceNow is a powerful platform, which is intended to manage everything as a service, enables the modern organization to function more quickly and flexibly than ever before. It accomplishes this by focusing on the activities, tasks, and processes used to create a typical workday via a service-oriented lens.

Important About CSA Exam

It's important to know the CSA Exam purpose, testing options, examination content coverage, test framework, and the prerequisites necessary to become a ServiceNow Certified System Administrator.

The ServiceNow System Administrator Certification demonstrates that a successful candidate has the skills and essential knowledge to contribute to the configuration, implementation, and maintenance of the ServiceNow system. Successfully passing this Certification exam also establishes a set of skills necessary to continue in the ServiceNow Certification paths. It is a prerequisite for advanced courses.

Successful candidates have system administration roles and belong to groups that allow administrative access to ServiceNow administrative applications and modules.

Prerequisites

This exam does not have any specific requirements such as familiarity with programming languages such as JavaScript or C++.

A successful candidate can;

- Create new applications and new modules to establish an information hierarchy.

- Manage users, groups, and roles.

- Personalize and create forms and fields for the various roles and groups to target company requirements.

- Define Service Level Agreements (SLAs), notifications, and reports.

- Interact with the ServiceNow Knowledge Base for company use.

- Implement Application Security by using high security settings and access controls.

- Populate the Configuration Management Database (CMDB) by defining Configuration Items (CIs) for company use.

- Move data in and out of an instance using update sets, import sets, and transform maps.

- Build Service Catalog items and variables, and apply workflows

CSA Exam Structure

The exam consists of 60 questions delivered in a 90-minute period. The following table shows the knowledge domains measured by this exam and the percentage of questions represented in each domain.

Multiple Choice (single answer)

For each multiple choice question on the examination, there are multiple possible responses. The person taking the exam reviews the response options and selects the most correct answer to the question.

True or False

An examinee is presented with a statement and is asked to select the correct answer from the two options; the statement is either true or false.

Matching

An examinee is presented a list of items and is asked to match each item it to its correlating item displayed in a separate list.

Testing Process

Each candidate must register for the exam. During the registration process, each test taker has the option of taking the exam at an Authorized Testing Center or as an online-proctored exam. In both testing venues, the Certified System Administrator exam is done through a consistent, friendly, user interface customized for ServiceNow tests.

The Kryterion Testing Network is worldwide and all locations offer a secure, comfortable testing environment. Candidates register for the exam at an exact date and time so there is no waiting and a seat is reserved in the testing center

Each candidate can also choose to take the exam as an online-proctored exam. This testing environment allows a candidate to take the test on his or her own system. Access to a web browser, a webcam, and broadband access to the Internet is required.

Exam Results

After completing and submitting the exam, results are immediately calculated and displayed to the candidate. A Pass or Fail message is displayed, giving the candidate immediate feedback.

NOTE: Actual scoring information is not provided to protect the integrity of the exam.

Module 1

System Information Architecture

About ServiceNow

Application Platform-as-a-Service (aPaaS)

- Configurable Web-Based User Interface

- Easy-to-use Services Management Solution

- True Cloud with High Availability Architecture

- Common Data Model with Flexible Table Structure

- Multi-Instance Architecture based on Single System of Record

- Modern Enterprise Cloud Computing Model with Strong Foundation

- Single System of Engagement i.e. Secure, Compliant, Scalable and Non-Stop

- An application service is a set of interconnected applications and hosts configured to offer a service to the organization

ServiceNow (Now Platform) Capabilities Out-of-the-Box

- Standardize the Service Delivery

- Automate the manual repeatable processes

- Takes care of all application infrastructure requirements

- Each ServiceNow solution provides its own guided setup

- Consolidate the organization's business processes at one place

- Allows to focus on building, running, and managing great applications

- Deliver Fast, Cut Cost and Boost Satisfaction to focus on core business

- Helps to Predict, Digitize, Optimize, Monitor and Resolve Security Risks

- Prebuilt Libraries, Tools, Templates, Scripts, Macros, Integrations, backups, and security

- Organizations may have different instance environments for Development (Dev), Test, and Production (Prod)

- ServiceNow feature "Guided Application Creator" can be used to begin the creation of an application

- Click on the List Mechanic gear above Check Boxes will display slushbucket

- "Retroactive Start" sets the start time equal to when an Incident was created

- Different form templates can be applied to a form by accessing the "Template Bar"

- There are two ways to add an application to a ServiceNow instance for development

  o Use studio to import an application from source control

  o Use the guided app creator to create an application

Resourceful Now Learning (Self-Learning)

- The learning sources of ServiceNow

- Product Documentations

- Knowledge Base

- TechBytes Podcast

- ServiceNow Community

- Resources to learn more about ServiceNow, includes ServiceNow website, documentation, support, Now community, Idea Portal, Interactive Space, Now Creators and more!

- servicenow.com

- docs.servicenow.com

- support.servicenow.com

- "ServiceNow Community" is leveraging an interactive space for customers, partners, and employees to connect, collaborate and engage to share their knowledge about ServiceNow products and services

- You have complete control to determine personal information displays publicly on your Now Profile. The only required information is a Display Name and First Name of your choosing

- Now Learning provides targeted courses, certifications, and simulators by Role, Level, Product, or Other Interests and also tracks your progress towards completing your personalized learning paths

- Now Creator profile URL can be used into job postings, websites, and emails to share your success stories and achievements with the world

- The Idea Portal allows you to collaborate and vote on top ideas for ServiceNow product enhancements

- Free ServiceNow developer instance to practice and play around new ideas

- "Developer Portal" supports the efforts and growth of ServiceNow developer community with self-paced trainings and documentation

- developer.servicenow.com

- Support available on the customer support portal, also known as HI (High Incident)

- You can also access ServiceNow corporate social media resources on LinkedIn, Twitter, YouTube, and Facebook.

ServiceNow Mobile Apps Major Features

- Global search to find people, service, items, and articles

- Push notifications for access to important information

- Submit, view, and update requests, issues, and tasks

- Offers two persona-focused ServiceNow Mobile apps

    o ServiceNow Mobile Agent

    o Now Mobile

Application Domains

- Global Domain: All users have access to see and manage records with permission

- Other Domains: Access is limited to Specific Users or Assigned Users, only users who belong to a domain can see domain specific records

- Types of ServiceNow releases

    o Feature

    o Patch

    o HotFix

ServiceNow is NOT

- Multi-tenant system with a risk to mingle each other data

- SaaS is Software as a Service

User Interface

(Forms, Lists, Context Menu, Branding Editor, CSS / Custom Style Sheets, UI Policy and UI Policy Actions, Filters, Wildcards, Homepage)

- Now Platform user interface is the primary way to interact with the applications and information in a ServiceNow instance

- Select Gear to System Settings are used to enable or disable form tabs in the platform

- personalize the system settings of the Now Platform user interface

- Cloning of an instance is used to copy production instance over a sub-production instance

- Home icon takes you to the homepage

- Fields and records are the components of a form

Now Platform Interfaces

- Next Experience Unified Navigation

- Now Mobile App

- Service Portal

Next Experience Unified Navigation

- This is the primary way to interact with applications and information in a ServiceNow instance which are also known as the UI component of the Homepage, includes

  - Banner Frame

  - Content Frame

  - All Menus (All, Favorites, and, History)

  - Application Navigator

- The function of the "Navigator" is to provide links to all applications and modules

- Application Navigation Search history is NOT stored anywhere

- Content type "UI pages" is NOT tracked in the history tab within the Application Navigator

- "Application" icon should you double-click to expand and collapse the list of all Applications and Modules

- Application Navigation (left nav) is populated based on the role of the logged in user

Impersonate User

- Impersonation option exist in "User Menu" which is used to become another user for testing purpose or to view the other user's environment

- Impersonation option is not visible in the mobile view of the platform

- Users with impersonator role cannot impersonate admin users

Types of UI Actions

- Form links (Related Links in a form)

- Form buttons

- Form context menu items

Save, Submit, Update, Insert, Insert and Stay

- **"Submit"** create a new record and return to the previously viewed page

- **"Save"** save the changes on an existing (old) record on the Form and keeps the Form open

- **"Update"** update changes on an existing (old) record and return to previously viewed page

- **"Insert"** or **"Insert and Stay"** both used to save a copy of the existing record instead of updating the existing (old) record

- **"Insert"** make a copy of the record and leave the form

- **"Insert and Stay"** make a copy of the record and stays on the form

	Create New Record	Update Old Record	Create a Copy of Old Record	Stay on Same Page/Form	Leave the Form/Page
Submit	X				X
Save		X		X	
Update		X			X
Insert			X		X
Insert and Stay			X	X	

History Tab/Menu

- History entries are created for many types of content including lists, records, and homepages

- By default, the History menu shows the past thirty history entries

- History entries are listed chronologically and divided into time periods

- System Administrator can configure a system property to change the default value of history entries up or down

- Selecting record from History tab is one of the quickest ways to navigate back to same record

- To track all UPGRADES made to system (System Diagnostics →Upgrade History)

- The History menu has got to be one of the most underutilized, underrated features when navigating the platform

- Reference for the System Property related to History → glide.ui.nav.history_length

Work/Activity Notes

- The ways to collaborate with team members on the Tasks

  - Activity Stream

  - Work Notes

  - User Presence

  - Connect Chat

- Activity Stream allows users to see a time-stamped history of all actions taken within a record

- "Notes" tab contains the activity stream for a task

- Activities are read-only entries made when records are created or updated

- "Work Notes" are for internal conversation for those resolving the task while "Additional Comments" are for customer visible comments

- "User Presence" feature facilitates synchronous collaboration and allow you to see who are online, their current status, and what they are viewing or editing, all in real-time

Context Menu

- Column Context Menu can be used to configure the list

- Lists provide context menus at three different levels

- List title menu

- List column context menu

- List fields context menu

- Form Context Menu provides options related to viewing, configuring, creating favorites and saving form data

- Different ways to create a favorites

- Select the star of the corresponding application or module

- Drag an individual record to the Favorites tab

- Drag the breadcrumbs of a filtered list to the Favorites tab.

- Users can access Column Context Menu to display actions that involve creating quick reports, configuring lists and exporting data

UI Policies

- UI Policy is a rule that applies to a form to dynamically change form information on the form itself

- UI policy action is configured to instruct fields how to behave on a form when a UI policy is triggered

- UI Policy action cannot delete a field

- The attributes which can be changed through UI Policy actions

  - Visible/hidden

  - Mandatory

  - Rad-only

- Data Policy enforce data consistency across the application

- When applying a data policy to the list view to make a field read-only, the system behavior is that field will appear editable, but the update will fail

- UI Actions can execute both client-side and server-side

- Data Policy can be run as a UI policy client-side

Form Designer

- By using "Field Navigator" on "Form Designer" Interface, a filed can be added to table and form

- To create a field on a table with Form Designer, from the Field Types tab, drag the field type onto the form and configure the Field

- Form designer was introduced after form layout to help design forms based on how they appear visually

- Form Layout needs to be configured to create a new form view

- When using Form Designer, the icons presented on each field include

  o Handle icon

  o Remove icon

  o Edit icon

Form Designer

- Branding Editor is used to preview the changes as they are made and applies a color theme to a portal

- To change the color of the Instance, use "Navigation Menu" under the "Content Management"

- The main Branding features which can be configured

  o Navigator background and text colors

  o Browser tab title and

  o System date/time formatting

- o Banner image, text and colors

- Some major benefits of branding the instance

  - o Accelerate adoption rates.

  - o Gain user trust

  - o Create a shared identity

- While implementing IT Service Management (ITSM), the ways to navigate to update Now Platform user interface branding includes the company logo and colors etc.

  - o System Properties → Basic Configuration UI16

  - o Guided setup → ITSM Guided setup

List View

- To configure a list, you can right-click on any column header and select Configure > List Layout. You can also access these options by clicking on the Column Options Menu

- The add and remove buttons will add the fields you want to the view and remove the fields you don't

- As shown below, the list collector interface allows you to select multiple items from an Available list and remove items from a selected list

- We can use the Move-up and Move-down buttons. Using these buttons will establish the order in which the fields will appear in your list view

Filters

- The list interface consists of a title bar, filters and breadcrumbs, columns of data, and search capability.

- A filter is a set of conditions applied to a table to help you find and work with a subset of data. In the Now Platform, filter conditions are also referred to as Breadcrumbs

- Breadcrumbs is the link that appear as underlined text to display content criteria

- Filters are defined with attributes i.e. Field Name, Operator and Value

- Possible results when entering 'service' in the Navigation filter

  o Any modules with names containing 'service'

  o All modules and sections or modules within the service desk application

- To apply a saved filter, click on the table name at the top of a list and select filter

- Use Filter Navigator component to see which Applications and Modules contain the word 'Task'

- Click arrow next to breadcrumb in list to change the Smart Filter Condition Builder

- Lists and Filters give you the power to sort, analyze, and view information in the Now Platform.

- Lists and filters help manage the records in a ServiceNow table.

- A list displays a set of records from a table. Users can search, sort, filter, and edit data in lists.

CSS (Custom Style Sheets)

- System Properties → CSS Properties used to change the banner, colors on your interface including homepage

- System Properties CSS Properties used to change the banner, colors on your interface including homepage, the two base tables of ServiceNow

  o Task

  o CI

Wildcard

- =Mysearchitem is the syntax to return the items that equal the phrase

- *(Asterisk) is used to search for values that contain a search term in a list

- Wildcard characters are the feature that can you use to narrow down search results in ServiceNow

- abc% is the syntax to return the items that start with abc

Natural Language Query

- The Natural Language Query allows you to filter the list data using natural language, instead of the condition builder. As you type, auto-suggestions for text will appear so you can click the suggestion before typing the complete filter.

- The Natural Language Query can be activated by installing the Natural Language Query (com.snc.nlq) and Predictive Intelligence (com.glide.platform_ml) plugins. Administrators will be able to turn this feature off by using the list control.

Work Flows

(Flow Designer, Process Automation)

Flow Designer

- Flows automate business logic for a particular application or process such as approvals, tasks, notifications, and record operations

- Prior to the introduction of Flow Designer, ServiceNow repeatable processes were created using a tool called Workflow Editor

- Don't use Flow Designer if existing logic already developed using ServiceNow workflow editor

- Flow Designer is Natural-Language descriptions of flow logic and a non-technical interface for building and enabling process automation capabilities, known as flows.

- Flow Designer is accessed by navigating to All > Process Automation > Flow Designer

Workflow Context

- When workflow invoked it loaded in the workflow context

- When workflow invoked it loaded in the workflow context

- Workflow versions captured in wf_workflow_version

- Connectors are not the part of Workflow

- Multiple version of the workflow can exist at the same time

- Only one version of the workflow exists in runtime for new contexts

- You can define visibility of variables on task form through workflow when it is created

- Business needs which can be addressed using Flow Designer are

    o Integration with 3rd party systems/tools

    o Orchestrate business processes across services

    o Reduce scripting to simplify upgrades and deployments

    o Extend flow content by subscribing to IntegrationHub or installing spokes

    o Promotes process automation by enabling subject matter expert to develop and share reusable actions

- The roles which controls Flow Designer functions are

    o flow_designer

    o flow_operator

- o action_designer

- Types of Flow Designer triggers

  - o Record-based

  - o Schedule-based

  - o Application-based

- Flow Designer provides

  - o A single environment or interface to build and visualize business processes

  - o Quicker flows or processes (time-saving) without using code

  - o Configuration and runtime information to create, operate and troubleshoot flows

- The applications delivered by ServiceNow are divided into four different workflows:

  - o IT Workflows

  - o Employee Workflows

  - o Customer Workflows

  - o Creator Workflows

- Actions are the operations executed by Flow Designer, such as looking up a record, updating a field value, requesting an approval, or logging a value

- Application-based type of Flow Designer trigger is needed for an Inbound Email Action

- Follow-up drilling or actions in Flow Designer

  - Look up and delete multiple existing records as a single action

  - Duplicate actions or subflows within a flow

  - Test a flow, subflow, or action in the background

Process Automation

- To use Process Automation Designer, navigate to All > System Definition > Plugins, then search for and Install the Process Automation Designer plugin

- Process owners are able to author cross-enterprise workflows within a single unified interface connecting multiple flows and actions, consolidating business processes across the organization

- Processes documented in the Process Automation Designer, are called Playbooks

- Process Automation Designer Benefits includes

  - Guiding end users to complete a process in a task-oriented interface

  - Consistent record lifecycle from beginning to end to add automation to the process

- o Passing data between activities and stages of business processes

- o Specifying the conditions and the order for activities and stages

- o Structured process design interface that makes it easy to define, visualize and manage activities and stages in a Kanban-style board

- Process Flow Formatter is used to display different stages in a Liner Process Flow

  - o Can be contained in a spoke

  - o Represent reusable operations for use across multiple flows

  - o Provide the ability to build your own custom actions

Events

- Event is an indication to the ServiceNow process that something notable has occurred

- Event means, log records indicating something notable has occurred in the system

- The Event queue contains a record of events

- In order to be used, an event record must be Registered

- Publish step makes workflows available to users

- Events can occur by

    o User actions

    o Scripts(Business Rules/Workflows)

- Events trigger Notifications

    o An email sent to the user when any event occurs in Servicenow

    o Event fired a notification

Triggers

- A trigger starts a flow when the conditions of the trigger are met

- A Record-based Designer Trigger is required to look up a record, update a field value and request approval

Data Pills

- Each time you add an Action to a flow, Flow Designer adds a data pill to store its output results

- Data pills are added to the Data panel, when you create Triggers and Actions in your workflows

- When a flow runs, it generates the data pill runtime value, which remains the same for the duration of the flow. For example, a data pill for [Trigger->Incident Record] always contains the incident record values from when the flow started

Process Fulfillment

- Fulfillers with permissions can create additional ad-hoc tasks for a requested item

- A fulfiller has completed all tasks for one of the items in a request, and the tasks are set to Closed. Now the requested item is automatically set to Closed Complete

- Below are some of the frequently used actions contained in the spoke ServiceNow Core

    o Ask for Approval

    o Create Record

    o Delete Record

    o Look Up Record

    o Wait for Condition

- Different ways to apply approvals

    o Approval Rules

    o Workflow

    o Process Guides

- Process of making catalog workflow

    o Define Workflow

    o Create Workflow Activities

- Workflow types does the Now Platform provide
    - Employee
    - Customer
    - Information Technology (IT)
- Stages of Workflow are
    - Waiting for Approval
    - Fulfillment
    - Delivery
    - Completed
    - Or Request canceled
- Activities in workflow
    - Approvals
    - Conditions
    - Notifications
    - Tasks
    - Timers
    - Utilities
- Basic components of the Workflow

- o Approvals

- o Notifications

- o Task

- Playbooks can be made visible in an Agent Workspace to outline the steps an agent should take for different scenarios. For example, this playbook, in the Customer Service Management workspace, is for handling customer complaints.

- The process is organized into Stages and Activities

    - o **Example Stages:** Assign, Create, Review and Update

    - o **Example Activities:** Simple Instruction, Wait for Interaction assignment, show Knowledge Article

Knowledge Management

- Knowledge management is an application which provides a centralized location for creating, categorizing, viewing, and governing information related to the flow of work

- The value of Knowledge Management increases as users interact with it

Knowledge Base

- Knowledge Base is a repository of important information that provides information to knowledge consumers

- Knowledge base hierarchy

  o Topics

  o Categories

  o Articles

Knowledge Articles

- Knowledge Base Articles are referred in NEWS module

- Available knowledge article types are

  o HTML

  o WIKI

- In a Knowledge base, articles are grouped according to categories

- Common examples of knowledge articles include policies, self-help tips, troubleshooting, and resolution steps

  o The main purpose of Embedded Help

  o Coach users on using a custom application

  o Display content based on user role

  o Display content based on query parameter values

  o Provide specific written or video-based instructions for a custom application

- Knowledge articles are tracked all the way from draft through retirement

Publishing Knowledge Articles

- The ways, Knowledge base can be populated with knowledge articles

  o Manually (Creating articles directly in the ServiceNow platform)

  o Integrate (integrating with a WebDAV compliant source)

  o Import articles (Microsoft Word files)

- From the Service Catalog with a Record Producer

- Create articles from cases, events, existing incidents or tasks

- Knowledge article authoring rules using Microsoft Word

  - You can author and access knowledge articles in Microsoft Word by deploying the Knowledge Management - Add-in for Microsoft Word

  - The text size of a knowledge article created in Word must be less than 1 MB

  - It is only supported in the Word online applications and not in the Word Desktop app

- The sequence of publishing Knowledge articles is Draft→Review→Publish

- Once a knowledge article is saved as a draft, you can follow its progress in the Knowledge list

- If you select Publish when creating articles manually or by import, It trigger the publish workflow assigned to the knowledge base

- This is mandatory to select "Knowledge Base" when importing Microsoft Word documents as knowledge articles

- The best way to share a knowledge article with another user is a permalink

- A record is added to table (Knowledge Use [kb_use]) each time a knowledge article is viewed

- Flagging article is a good way to privately suggest an article revision to the knowledge manager from the Service Portal

- When opening a knowledge base from the service portal, sorting options are available by default on the knowledge article list Views

    o Newest

    o Alphabetical

Accessing and Managing Knowledge Articles

- User Criteria is used in Knowledge Management to control who can read and contribute at different levels

    o Knowledge Article

    o Knowledge Block

    o Knowledge Base

- User Criteria used to define conditions that determine which user can create, read, write and retire knowledge base article

- Only Knowledge Base Managers able to contribute if no User Criteria is specified for 'Can Contribute' in a Knowledge Base

- To allow users to access Knowledge Base without logging in, make it public and protect with role

- On the homepage, Catalog Item will access Knowledge

- By collecting knowledge usage metrics you can measure and track article views, page views, and searches performed by unauthenticated users

- Periodically reviewing documents identified as "Most Useful" is a great way to stay informed

- Knowledge articles are tracked all the way from draft through retirement

- The ways, an end-user can leave feedback about an article

  - Leaving feedback/comment about an article

  - Give a 1-5 star rating

  - Marking the article as helpful or not helpful

Knowledge-Centered Service (KCS)

- Knowledge-Centered Service configuration (KCS) is a method for the creation and continuous improvement of knowledge based on the experience of agents and the patterns observed by knowledge reuse

- If Knowledge-Centered Service (KCS) is used in your organization, agents and resolvers can create knowledge articles directly from cases or incidents

- ServiceNow also has the capability to integrate with WebDAV compliant knowledge tools, such as SharePoint

Virtual Agent Support

- Virtual Agent offers a web-based interface available for Service Portal, iOS and Android mobile environments

- Virtual Agent can be used in conjunction with Knowledge Management It is another way to interject knowledge right into the process

- ServiceNow's Virtual Agent is a Conversational Bot Platform that is the first stop before seeking live agent support, provides assistance to help users obtain information, make decisions, and perform common work tasks

- Virtual Agent also supports third-party messaging applications through ServiceNow adapters for Slack, Workplace, and Microsoft Teams

- At any time, users have the option to switch to speak with a human agent for assistance

- "Topic Blocks" feature creates reusable components to run common tasks in Virtual Agent conversations

- Virtual Agent Benefits

   o Accurate information and quick assistance, for common questions and concerns

   o Fastest and smoothest route always through a human agent

- o The biggest benefit to implementing Virtual Agent is your users can get immediate help, day or night

- o Virtual Agent could track down possible resolutions for you

- Virtual Agent offers a personalized customer experience by automating typical Tier 1 support tasks to be accomplished, including

  - o Providing tutorial ("how to") information

  - o Querying or updating records (for example, get the status on cases or incidents)

  - o Gathering data, such as attachments, for the agent

  - o Performing diagnostics

  - o Resolving multi-step problems

- The typical Tier 2 support tasks that virtual agents can perform so that the support agents can focus on more complex user issues

  - o Answering FAQs

  - o Providing how - to information

  - o Performing diagnostics

- The components that are working behind the online support

- o Knowledge Bases

- o Service Catalog

- o System Records

Module 2

Administer Users & Define Application Security

Roles and Rules, Users and Groups

Roles

- "Role" defines your capabilities in the application and modules

- A Role "as a collection of permissions and security rights" is used to control access feature

- Roles can contain other roles, a role inherit all the permissions of any other roles it includes

- Multiple roles can be assigned to a single user or a group of users

- Best practice is that rather than assigning roles to individual users, add the users to a group and assign the role to the group, avoid assigning roles directly to individual users

- By assigning a role containing other roles to a group, all users in the group are granted access to the permissions of any contained roles

- When new roles are assigned, the user would need to log out and log back in before the new permissions take effect

- ESS have NO Role

- ESS Roles allow users to view knowledge articles without logging in

- No special role is required to access the template bar or create templates

- The roles which cannot be delegated

  - Admin

  - Role_delegator

- view_changer role can switch their form view

- Administrators (admin) role can manage all aspects of the Service Catalog application and the scripting functions

- ITIL is the minimum role required to create incident templates so you can create incidents for similar issues quickly

- The modules which End User can see

  - My requests

  - Requested items

- Social Q&A feature allows end users to post questions and answers

- The users who can add or remove widgets in a dashboard

  - Dashboard owner

  - Users with edit access

- o Dashboard admins

- Non-admin users cannot add a user to a group that contains the admin role

- To grant the admin role to a user, the granting user must also have the admin role

- Built-in roles provided by ServiceNow

    - o System Administrator (System Admin)

    - o Specialized Administrator (example: catalog_admin)

    - o ITIL(itil)

    - o Employee Self-Service

    - o Approvers (approver_user)

Groups

- A user is an individual that has been granted access to ServiceNow instance

- A group is a set of users who share a common purpose or need access to similar information for various purposes, such as approving change requests, resolving incidents, receiving email notifications, or administering the Service Catalog

- Groups organize users into sets for easy maintenance

- A user can be removed from one group and added to a new group

- Users working in ServiceNow are typically assigned to one or more groups

- The best practice is, assign users to groups and map roles to groups

- Apply general roles to large groups and specific roles to smaller groups

- The users are related to roles and groups as "One-to-many relationships"

- "My Groups Work" is the module displays a list of tasks assigned to a user's group but not yet assigned to an individual user

- Group types which are provided in the base system

    o Catalog

    o ITIL

    o Survey

- Following deployment permissions can be assigned to a group or user for a specific application when using Manage Developers

    o Upgrade App

    o Publish to App Store

- o Manage Update Set

- The permissions which can be assigned to users in groups

    - o Approve, change, or resolve incidents and requests

    - o Receive email notifications

    - o Provide a reference for alerts and notifications

Self-service

- Self-Service applications are available to all users

- self-service users are the users without any assigned role permissions can still log in to ServiceNow and access common actions, such as viewing a homepage, accessing the Service Catalog, viewing knowledge articles, and taking surveys

- The actions which Self-Service User can assess when logging into ServiceNow

    - o Viewing a homepage

    - o Taking surveys

    - o Viewing knowledge articles

Delegated Development

- The features of the delegated development and deployment

- o Application Collaboration

- o Allow delegated developers to delete applications

- o Assign source control permissions

- Delegated Development granted to non-administrators in order for them to be able to develop applications

- "Delegate", receives notifications, meeting invites on the behalf of other users

- The "Subscriptions" feature allows users to manage notifications they receive about various

- Following are the notification preferences that users can manage

  - o Disable receipt of specific notifications

  - o Apply a schedule to a notification

  - o Enable/disable notifications

Security

- A security_admin privilege is created, after the High Security plugin is activated

- The elevated role of Security_admin is required to create and edit Access Control rules

- A user with only the admin role cannot grant the security_admin role to other users

- To provide added security, no user can have both of the explicit roles. (snc_internal & snc_external)

- Limit of file attachment can be set through Security → System Security

- Plug-Ins cannot be removed, but they can be disabled

- Remove pagination count to speed up loading large lists

- A mapped graphic image on a homepage that are packed reports called a gauge

- Page loading time of the homepage can be improved by reducing the number of gauges

- It is crucial to protect sensitive data. Realize not every member of your organization needs access to all information at all times.

- To help improve ServiceNow instance performance, system administrators can set the glide.ui.per_page system property to limit the paging control options available to users. These options determine the number of list rows displayed per page

- Locate failed changes and associate incidents to facilitate quick analysis of impact and to reduce or eliminate downtime

- SYS ID is another name for the 32-character GUID

Authentication

- The Now Platform can support many different methods of user authentication

- Adaptive Authentication enables you to create polices based on a criteria-like role, IP Address, or Group

- Authentication method OpenID Connect is used as a Single Sign-On (SSO) identity provider (IdP) to allow users to log in to ServiceNow using their social identity provider, like Google and Okta

- User authentication method "External Single Sign-on (SSO)" is used when it authenticates the user name and password configured in identity providers with a matching user account in the ServiceNow instance

Access Control List (ACL)

- ACL is the part of Contextual Security

- ACL provides permission to access any object in ServiceNow

- ACL is the list of all Access Controls for a table

- ACL is applied on Table, as a security rule defined and set at the row and column level

- Elevated privilege: A role that has special permission for the duration of the log in session

- "Session" is an elevated privileges

- An elevated privilege is the role that has special permissions for the duration of the log in session

- Elevate Roles prevents accidents caused by forgetting the power of owning an important role, such as security_admin. It requires you to manually accept that responsibility before using it

- Elevate Roles provides an extra layer of precaution for powerful roles. Administrators often use it when modifying high security settings

- Through setting on the List Control page, System Administrators can speed up loading large lists by removing the calculation of the total number of records (pagination count) on list control page

- If there are row and field ACL's, User has to stratify both to have access to the field

- The order by which access controls evaluated

  - First at the Table-level (most specific to most general)

  - Then at the Field-level (most specific to most general)

- Access control list rules specifies

  - The permissions required to access the object

  - The object and operation being secured

- The three (3) levels of Access

  - System (username/pw)

  - Applications and Modules (Roles)

  - Tables and Fields (System Properties/ Access Control)

- The record operations that can be secured by ACL rules

  - Personalize_choices

  - Save_as_template

  - Edit_task_relations

- The types of permissions that can be configured or defined in an access control rule and must evaluate to True to grant access to a resource

    o Scripts (a script that sets the 'answer' variable to true)

    o Condition Expressions

    o The user must have one of the roles in the required roles list. If the list is empty, this condition evaluates to true

- Order of Evaluation of ACL

    o The condition must evaluate to true

    o The script must evaluate to true

    o The user must have role

    o The other ACL matching rules must be true

Module 3

Monitor, Communicate and Report on Activity

Business Rules

(Server Script, Client Script)

- Business Rules apply Globally

- Business rules do not monitor fields on a form

- Business Rule is a control that applies permissions, sends notifications and triggers other processes when a record is displayed, inserted, updated or deleted or when a table is queried

- "Order" in a Business Rule represents Order in which it will be executed

- A Business Rule Condition field return value "True" if the field is empty

- Business Rule is a piece of Javascript configured to run when record is displayed, inserted, deleted, or queried

- Business Rules and Data policies run on server side

- Business Rule script runs when a record is displayed, inserted, updated, deleted or when a table is queried

- Business rules stored sys_script table

- Script Include is a Server Side Script and it is a Reusable Code

- To run Data Policy on the Client Side instead of Server Side (Select the checkbox "Use as UI Policy on the Client"

- UI Policy and Client scripts run on client side

- We can't write schedule job for "Form Validation at client side every day"

- Code Search is used to search code across all the objects in your application

- The best response times are

  o Total Response Time - network browser and server 3 seconds

  o Server Response time below 800 ms

- Client-side scripts execute on web browser

- JavaScript is the main language used for scripting in ServiceNow

- We can't trigger an email notification using Client Script

- The variables that can't be used on the Client Script

  o Procedure.variable

  o gs.include()

- The variables that can be used on the Client Script are

  o Jslog

- o g_form

- onSubmit Client Scripts execute their script logic in following conditions

    - o When a user clicks the save menu item in the additional actions menu

    - o When a user clicks the update button

    - o When a user clicks the Submit button

- SHOULDN'T use client scripts when you can perform the same actions use UI policies or Access Control Rules

- Client scripts only apply when accessed through the form

- Following are the four types or ways of Client Scripts executed

    - o OnLoad()

    - o OnChange()

    - o OnSubmit()

    - o OnCellEdit()

- onChange client script runs on change in the value of a particular field

Service Level Agreement (SLA)

- Service Level Agreements is the functionality used to

    o Track the amount of time a task has been open

    o To ensure that tasks are completed within an allotted time

- SLAs are stored in table u_sla and task_sla

- SLA is a measurement of the set amount of time for a task to reach a certain condition

- SLA's include actions that can be triggered anytime

- When an SLA task does not reach the condition or does not matched called "SAL Breach" or "SLA Breached"

- SLA, OLA and UC are configured in SLM module

    o SLA - Service Level Agreement

    o OLA - Operation Level Agreement (how departments work together)

    o UC - Underpinning Contracts (outside suppliers)

- Following are the conditions/steps defined on SLA which can trigger SLA

    o Start condition

    o Pause condition

- o Stop condition

- Pause condition for Incident SLA presents awaiting<something>

- Major components that empower SLA

  - o SLA Definition

  - o Task SLA

  - o SLA Workflow

  - o SLA Automation (Business Rule and scheduled job)

- Database view type of relationship is recommended between Incident and SLA tables to report on incidents resolved by SLA per incident category

Notifications

- System Notification → Email → Notifications (contains the notification which are currently defined in ServiceNow)

- Notifications inform users of important events, such as task resolution

- System Settings In Auto Opt-in present notification

- Notifications in the Now Platform occur through the following methods

  o Emails

  o SMS

  o Meeting invitations.

- Following are the main sections when you are configuring email notifications

  o When to send

  o Who will receive

  o What it will contain

- "Who will receive" is the section where the email notification configuration will be define

- "When to send" tab within the email notification configuration, you identify that a record insert or update triggers a notification

- Schedule a Job or Modifying existing record is not possible via an inbound actions but sending an email notifications

- We re schedule jobs to update the old records, email notifications and to check conditions

- If a request is rejected then a notification is sent and sets the status to cancelled

- Sending an email notification is possible via an inbound action

Reports

- Possible source types to build the reports

    o Data source

    o Table

- "Drilldown" is a report feature to display a subset of the report

- A homepage is a dashboard of frequently used content with reports

- "Embedded PNG" is used to embed report visualizations in the body of scheduled report emails

- Analytics Q&A application/functionality allows you to generate a report by entering a query instead of going through the full Report Designer menu

- "Chart" field data type adds a report to a form

- You can use Report Designer to configure reports, which allows you to choose colors, title and chart properties

- Following are types of Standard reports visualization that can be generated from a list of records

    o Pie Chart

    o Bar Chart

    o List

- o Calendar

- o Pivot Table

- o Box Chart

- o Trend

- Pareto Chart can be used to represent individual values as bars and cumulative total as line

- Use aggregate field values when creating and sharing visualized reports on metrics in the Report Designer

- By default, a report is shared with the report creator only

- "Select Share" is the recommended way to share a report

- The valid report sharing options are

  - o Export to PDF

  - o Publish

  - o Add to dashboard

- Performance Analytics (PA) allows users to create dashboards with widgets to visualize data over time in order to identify areas of improvement

- Reports need to be explicitly shared with either everyone, or specific Groups and/or Users so they can be viewed either in the Reports application or on a dashboard.

- It's must to save the report after modifying its sharing settings

Visual Task Board

- **"Freeform"** is the type of Visual Task Board that CANNOT be built from a record list

- **"Guided"** Visual Task Board type automatically updates the tasks when the respective cards are edited or change lanes

- **"Flexible"**, based on the field values from record lists, Records do not update when cards are moved

- **"Owner of the Board"** is the one who has access to create, edit or delete sorting criteria in Visual Task Boards

- **"Compact Cards"** allows you to see more information on the same screen when using Visual Task Boards

- **"Lane Filter"** in Visual Task Board enable you to create a separate lane for the cards that contain empty values, such as cards with no assignees

- Tasks can be described as

    o Tasks minimize the possibility of human error

    o Tasks are repeatable processes

    o Tasks lead to quicker resolution times

Module 4

Implement Self-Service and
Automation

Service Portal

- The Service Portal provides a user-friendly self-service experience, by providing access to specific features, using widgets. Users are able to

    o Search for articles, catalog items, records

    o Submit requests

    o Browse the corporate news feed

    o And much more!

- Statements about the Portal

    o Containers can be fluid layout

    o Viewport size changes when a page is resized

    o Page Layouts are responsive to screen resolution

    o Page Layouts are responsive to Device Type

- The Service Portal homepage can be accessed by navigating to https://<instancename>.service-now.com/sp

- Both expert developers and beginners with the proper permissions can configure portals to create engaging experiences

    o Less technical users can make basic configuration changes to the UI, using

Branding Editor and other components of Service Portal.

o More advanced users can edit and extend portals, pages, and widgets.

o Expert users can use the Widget Editor tool to write scripts to power a portal and even create rich web applications on the Now Platform

Service Catalog

- Service Catalog is used to digitize the flow of work across the enterprise

- Service Catalogue contains a collection of orderable products and services

- Service Catalog is a robust ordering system for services, hardware and software and the central repository of goods and services that an IT service desk provides for users

- Service catalogue component "Order guide" allows for multiple catalogue items to be logically grouped as one request

- Following are the major components of Service catalog

  o Record Producers

  o Items

  o Variables

  o Order guides

  o Workflows

- Service catalog displays

  o Catalog items

  o Order Guides

- Record Procedure

- A variable set is a unit of variable which can be shared with catalog items

- REQ→RITM→TASK catalog form and click on Order now button to create request

- Service Catalog can be accessed "All > Self-Service > Service Catalog" using from the Next Experience Unified Navigation

- To approve a catalog request, right click request record in request list and click approve

- User Criteria [user_criteria] contains the conditions that which user can access Catalog Items

- The order field in service catalog displays the catalog item in the ascending order of this value

- Modify the "Order" field to change the sort sequence of fields or tasks

- Catalog Builder is used to create or edit(maintain) catalog items, through visual guided experience

- By using the Catalog Builder you view how a catalogue item appears in a conversational interface and modify the item if required

- While creating templates using Catalog Builder, you can specify values or restrictions for items created using the template (e.g. categories, variable types, and portal settings)

- Using the Catalog Builder you can

    - Create a catalog item and item templates

    - View the available catalog items and item templates

    - View catalog items that are recently updated

    - View the configured content that describes the catalog building process in your organization

- Variable type "Requested For" should be used for the requester when requesting a catalogue item on behalf of another user

- Variable type "Attachment" should be used to allow uploading an attachment from a question in a catalogue item

- System Administrator navigate "Maintain Items" to edit a catalogue item

- Under Request Something using the Service Portal, end-users typically access the service catalogue

- If an end user wants to report an incident, can use "Can we help you" category in the Service Catalog

- From the end user perspective, Categories and Subcategories, is used to manage products and services in the service catalogue organized

- Service Catalog > Catalog Definitions > Maintain Items is used to navigate in the Now Platform to see a list of catalogue items

- The purpose of service catalog workflow is to

  - Drive request fulfillment

  - Generate and assign approval

  - Generate and assign task or run scripts

  - Make Sub flows

- Service catalog workflow can be attached by using any of the methods below

  - Manually on the catalog item form

  - Automatically, based on conditions

  - Automatically, if no other workflow is attached

- There are two options to define the fulfilment process for a service catalogue item

  - Flow

  - Workflow

- Following is created when an order is placed for a catalogue item

  - A Request (REQ) record

  - A Requested Item (RITM) record

- o one or more Service Catalog Task (SCTASK)

- Roles related to Service Catalog application

  - o Administrator

  - o Catalog Administrator

  - o Catalog Manager

  - o Catalog Editor

- Administrator can manage all aspects of the service catalog application including

  - o Categories

  - o Catalog item

  - o Catalogs

  - o Advanced functions such as Scripting or Creating Business Rules

- Catalog Administrator also can manage all aspects of service Catalog application expect scripting functions

- Catalog Managers can edit and update their assigned catalogs, they may also assign catalog editors and assign the catalog to different managers

- Catalog Editors can edit and update their assigned catalogs, but they cannot reassign the catalog manager

- Following are the various items that orders from Service Catalog generate

  - REQ

  - RITM

  - Catalog Task

  - Assignment Group

- Following are the Service Catalog best practices

  - Define an order guide

  - Group items in order guide

  - Use questions to present item options

  - Service catalog → Maintain Item

- Following are the purpose of order guide

  - A single initial screen, where the customer fills in some initial information

  - A set of selected catalog items based on conditions derived from the initial information

- Following fulfillers can navigate to find catalogue tasks assigned to them and their groups to fulfil requests

  - Service Desk > My Groups Work

  - Service Desk > My Work

Record Producer

- A record producer is a module of Service Catalog to create records from catalog view

- Record Producer is the type of service catalogue item which is used to create an incident or raise an HR case

- Service Catalog record producers implement simplified forms allowing users to create task-based records with minimal input

CMDB

- CMDB stand for Configuration Management Database

- CMDB is a series of tables that contains all the assets and business services controlled by a company

- The statements about CMDB Data Manager

    o It is a policy-driven framework for managing CI life cycle operations

    o It can create policies that govern CI life cycle stages, such as retirement and deletion.

    o It is a wizard-like tool that provides a comprehensive solution for managing CIs

- The CMDB is the authoritative source of following information

    o Support group

    o Contact

    o Owner

- CI stand for Configuration Item

- BMS stands for Business Service Map

- A Configuration Item (CI) is any component of an infrastructure that needs to be managed in order to deliver a product or services

- A Configuration Item (CI) is a tangible device or intangible dedicated software in the CMDB

- Following are the three key tables in the CMDB

  - Base Configuration Item [cmdb]

  - Configuration Item [cmdb_ci]

  - CI Relationship [cmdb_rel_ci]

- cmdb_ci is the core CI table to store the basic attributes of all the configuration items

- Cmdb_rel_ci is the CI relationship table and is used to store CI relationship data

- CI Relationship Editor helps users to see reasonable relationships between configuration items

- cmdb_read is the minimum role required to subscribe or unsubscribe a CI

- Types of Configuration items

  - Tangible (e.g. hardware, software, servers) entities

  - Intangible (e.g. business services, email) entities.

- Following methods are available to populate CMDB

  - Discovery

- o   Import Sets

- o   Integrate with external CMDB

- o   Web Services

- o   Help the helpdesk

- o   Manual Input

- Organizations fails to complete implementation using CMDB if

  - o   Ill-defined relationship

  - o   Unknown Cis

  - o   Inconsistent Data Quality

- CMDB is still operational after upgrades and deployments of new applications or integrations

- Any changes to CI relationships in the CMDB table changes automatically reflected in dynamic application services

- Following are the benefits of using Multisource CMDB

  - o   Visualize the source of attribute values for each discovery source and at the attribute level

  - o   Revert CMDB data integration from a specific discovery source

- o Control CI updates at the discovery source and CI attribute level.

- o List Benefits of CMDB is the cost savings

- Configuration Management Database (CMDB) is the database used to store configuration records throughout their lifecycle.

- Following IT challenges can be solved with the CMDB

  - o Regularly maintaining complex data for accuracy

  - o Consolidating disparate Configuration Item (CI) data into a single Configuration management database

  - o Make sense of data to drive decisions and services.

- Following scores are found on the CMDB health dashboard

  - o Completeness

  - o Correctness

  - o Compliance

- Fields and values in the CSDM framework

  - o Life Cycle Stage Status

  - o Life Cycle Stage

- Data Management plugins are

    o Data archiving

    o Database Rotations

    o Many to Many task relations

- Common Services Data Model (CSDM) consist of

    o CMDB core tables

    o Best practice for CMDB data modelling and Data Management

    o Guidance on service modelling

- **"Support Group"** is the field on a Configuration Item (CI) record that may be used to route Incidents to the appropriate group to resolve CI related incidents quickly

- **"Service Offering"** is the CI class that is not included in the search results for configuration items, affected CIs, or impacted services within an Incident record

- **"Principal Class"** to be set for a CI class in the CI Class Manager to display in the Configuration item field look-up in the change form

- Following can be used to create configuration items (CIs) in the CMDB

    o Creation via service catalogue fulfilment processes

- o Creation triggered by asset management

- o Creation via the change management process

- Incident field "Short description" is used to predict the Configuration Item and Service fields' value via Predictive Intelligence solution definitions

- Following is the best practice when configuring Assignment Rules or Predictive Intelligence

  - o Setting an Assignment group based on Category and Subcategory

  - o Setting an Assignment group and User based on Category and Subcategory.

  - o Setting a User based on Category and Subcategory

  - o Setting an Assignment group and User based on the Short description

- Teams is the related list that can be used use to extend and track the different types of groups assigned to a Configuration Item

- Business Service Map graphically displays the CIs that compose a business service and indicates status

- In BMS Map, CI is displayed in the middle of the MAP showing both upstream and downstream CIs

- Base CI can take Parent or Child CI depending on the relationship nature

- CI Class Manager is the feature that enables you to add related entries for Identification and Reconciliation Engine Identification (IRE) rules

- List CI examples

- Computers/Devices on the network

- software contracts and licenses

- Business services

- Model/standard across applications helps to track life cycle stages and stage statuses for CIs effectively

- NOW Platform feature, in conjunction with field normalization, does CMDB leverage to check a CI's uniqueness automatically

- A stakeholder's role in implementing and maintaining the CMDB is to determine the specific CI information required to support these capabilities and build organizational buy-in

- "Help the Helpdesk" is a tool that populates the CMDB about your windows computer

- "Configuration Management System" is a set of tools and databases used to manage an organization's configuration data

Import Set and Transform Map

- Transform Map provides a guide for moving data from import set

- Importing data into ServiceNow: Every import must have at least one Transform Map

- To do list before importing data into ServiceNow

    o Understanding the data you plan to import.

    o Decide which source data maps to which target fields

    o Decide what to do with incomplete or erroneous data

- Following are the possible data sources for importing data into ServiceNow

    o Http

    o Csv

    o Xml

    o JDBC

    o TXT

- Transform map tool is used to determine relationships between fields in an import set table and an existing table

- ServiceNow can import data from following external data source types when a valid transform map is available

    o LDAP

    o OIDC

    o REST

- Following are the steps for using Import Sets and Transform Maps

    o Load Data

    o Create Transform MAP

    o Run transform

- Import Set is used to import data from various data sources and map data into SN tables

- The statements about Import Sets

    o Creating an extremely large import set can cause delays and system outages.

    o You can import data from several different file formats or external data sources.

    o Import Sets cannot add data to encrypted fields.

Data Import and Data Export

- The file formats to export currently displayed records in a list

    o JSON

    o XML

    o CSV

- xls, xlsx , PDF and PNG can be viewed directly in the platform using the document viewer

- The steps to make easy Import

    o Download Excel Template

    o Upload Import Data

    o Finalize Import

- User to Group mapping is stored in sys_user_grmember

- Groups stored in Sys_user_group table

- User record stored in the User sys_user table

- Roles stored in Sys_user_role table

- sys_popup form view is displayed when clicking on the reference icon of a field in a form

- Customer Update table (sys_update_xml) is used to track changes in an update set

- The first-page results column in the Search Event [sys_search_event] table contain

  - Table names

  - Sys_ids

- If a user has the VIP field set to true on their User [sys_user] record. Now when that user is selected as the Caller on an incident... a 'VIP' decoration is displayed next to the caller field and the user's name is shown in red in the caller field.

Module 5

Configure & Maintain SN through Update Sets & Upgrades

Database

- Data is stored in a table

- Each column in a list interface corresponds to a field in the table. Each row corresponds to a record in the table.

- Metrics measure data over time to show past history

- Metrics are used to measure effectiveness of IT Service Management process

- A red-dashed line indicate invalid data or mis-typed word

- Data inconsistency correction can be done automated using Plugin

- A Form displays fields from one record and can be used to edit the record data

- "Can Write" is not a database setting on the Application Access section of a Table

- Scope protects applications by identifying and restricting access to available files and data

- "Data Source" should be used when creating a report from a dataset with pre-defined conditions

- Before adding a new field to a table and form, check all those existing table fields, you might find what you are looking for without having to add something new. Just verify that the purpose of the existing field matches your needs.

- Dot-Walking is the process of gathering information from a series of tables by expanding reference fields or to select fields from other related tables

Coalesce

- Coalescing on a field (or set of fields) means the field will be used as a unique key

- Utilize Coalesce field to make the field unique and to avoid duplicate data while importing

- Coalesce option allows you to update existing Target Table records when importing data from an import set

- If coalesce is not configured then import process treats all imported rows as new records and does not update existing records

Data Dictionary

- Data Dictionary stores structure and relationship definitions

- Data Dictionary in ServiceNow is used

    o To describe the database structure

    o To make the fields mandatory

    o To make a field unique

- The tables that provide Data Dictionary and Relationship Information

    o Sys_dictionary

    o sys_documentation

    o sys_db-object

Database Views

- The limitations of database view are

    o It is not possible to edit data within a database view

    o In a clone request, database view tables cannot be added as a data preserver

    o Database views cannot be created on tables that participate in table rotation

- The Incident Database views for the Service Management plugin in the base system

    o Incident_state

    o Incident_metric

    o incident_time_worked

- A "View" is used to call a saved version of a personalized form

- The view name is listed in brackets next to the record's display name in the form header

- If Prefix of Incident needs to be changed, System Definition → Number Maintenance (Number Maintenance Table), e.g. (to change prefix of number let's say from "PRB" to "PRBLM")

- Auto populated table name prefix as (u_) when creating a new custom table in the global application, If user creates a table "abc", ServiceNow will name it as u_abc

- "String" data type in a survey metric has the field validation option to validate email, IP address, phone number or URL, however, Long String is not a valid field type

- The different ways for ServiceNow tables to be related to each other

    o One-to-Many

    o Many-to-Many

    o Database Views

    o Extensions

- The different levels of ServiceNow security before an end-user has the capability to perform CRUD (Create, Read, Update, Delete) operations on a table

    o Database Access

- o User Authentication

- o Application and Modules Access

- Data Visualization Strategies are

    - o Reports

    - o Charts

    - o Dashboards

- The features to auto-assign all new Hardware category Incidents to a certain group

    - o Data lookup rules

    - o Assignment rules

    - o Business rules

- Tables and Fields can be viewed and manipulated at

    - o Record List

    - o Form

- Task [task] is a base table that is not extended but another table can be extended from it, like

    - o Problem

    - o Incident

    - o Change Request

- sys_id is the unique identifier for each record in the table

- You can delete the tables starting with "u_" permanently

- Inbound Email Actions: Setting the value for the field in a target table and sending email back to the source that triggered the action

- The Email in the user table is considered a column

- The possible outcomes when coalescing detects a match between a record in the staging table and a record in the target table

  o Keep the record already in the target table

  o Overwrite the record in the target table with source data

- Each record and each field correspond in a table as below

  o Each record corresponds to a row in a table

  o Each field corresponds to a column in a table

- The four access control rules that the system creates by default when a custom table is created

  o Create

  o Delete

- o Read

- o Write

- Both Tables or Tables & Columns modules can be used to create a new table

- The below created by default when you create a new table

  - o Application Menu with the same name as the table Label

  - o Module with the plural of the table Label

- The Schema Map is the Graphical Representation of Relationships between Tables

- The different Schema relationship types that are supported for tables

  - o Referenced by

  - o Referencing

  - o Extended by

  - o Extending

- Reference field type displays records from another table

- The interfaces for viewing and manipulating tables

  - o Record list view

- o Schema map

- o Tables and Columns module

- Access Control in Tables and Fields sets security for rows and columns

- Table security set by using System Property in Tables and Fields

- A List is called when you display the contents of a table

- Fields and Records are the components of a table

- List reports type requires access to the data to view it

- Dashboards can contain Widgets and Tabs

Field Types and Variables

- Following field types have a one-to-many relationship

    o Reference Field

    o Glide List

    o Document ID Field

- FX Currency is field type that allows to use positive and negative values

- Currency and Suggestion are the valid field type

- The fields that cannot be edited from List Editor

    o glide_time

    o html

    o Array fields

- g_form.showErrorBox ("Field_name", "Hello World") will put "Hello World" in an error message below the specified field

- Modify the order field to change the sort sequence of fields or task

- A field represents Column

- The three key attributes of every field

    o Field Name

- o Value

- o Field Label

- Modify the order field to change the sort sequence of fields or tasks

- The field status indicators are

  - o Light red - required but has a saved value.

  - o Green - Modified field content - need to save.

  - o Red - Required and needs value.

  - o Orange - Read-only

- Multiple Choice, Single Line Text and Select Box are the type of elements in ServiceNow

- A modular unit of variables that can be shared between catalog items

Integrations

- IntegrationHub can be used to extend the Flow Designer to call 3rd party systems such as automating Microsoft Services and infrastructure using PowerShell and REST

- ServiceNow product "integrationHub" provides the ability to integrate with 3rd party applications without scripting

- Scalable Flow Designer benefit corresponds to extending flow content by subscribing to IntegrationHub or installing spokes

- REST API Explorer is used to test API requests to a ServiceNow Instance

Update Set

- Customizations is another name for Customer Updates

- An update set is a group of customizations, personalization or changes that can be captured, packaged and moved from one instance to another instance

- Through update set, a group of one or more changes that can be moved from one instance to another altogether

- One update set should include many changes (instead of few)

- Two update sets can be merged

- Homepages and Content pages are not captured in an update set by default, but can be added to update set only by manually

- User customization records not upgraded

- Update_synch attribute identifies that it should be captured in update sets

- This is what automatically captured in an Update Set

    o Form

    o Table

    o Data

- Can be captured in an update set

  o Business Rules

  o Published Flows

  o Report Definitions

  o Fields

  o Roles

  o Views

- Cannot be customized with update sets

  o New Records

  o New users and groups

  o Modified data

  o Schedules

- Following is the sequence to run an Update Set on an instance

  o Retrieve

  o Preview

  o Commit

  o Apply

- Data Records and Dashboards are not captured in an update set by default

Glossary

Glossary

AJAX	(Asynchronous JavaScript And XML)
API	(Application Program Interface)
APM	(Application Portfolio Management)
AWS	(Amazon Web Service)
CAD	(Certified Application Developer)
CAS	(Certified Application Specialist)
CAS-PA	(Certified Application Specialist – Performance Analytics)
CCS	(Catalog Client Scripts)
CEs	(Underpinning Contracts)
CIs	(Configuration Items)
CIS	(Certified Implementation Specialist)
CMDB	(Configuration Management Database)
COE	(Center of Excellence)
CPG	(Cloud Provisioning and Governance)
CR	(Change Request)
CRUD	(Create, Read, Update, Delete)
CSA	(Certified System Administrator)
CSM	(Customer Services Management)
CSS	(Custom Style Sheets)
Disco	(Discovery)
EM	(Event Management)
FMA	(Multi-Factor Authentication)
FSM	(Field Service Management)

FTPS	(Secure File Transfer Protocol)
GRC	(Governance Risk and Compliance)
HAM	(Hardware Asset Management)
HRSD	(HR Service Delivery)
HTML	(Hyper Text Markup Language)
IaaS	(Infrastructure-as-a-Service)
IRE	(Identification and Reconciliation Engine)
ITAM	(IT Assets Management)
ITBM	(IT Business Management)
ITIL	(IT Infrastructure Library)
ITOM	(IT Operations Management)
ITSM	(IT Services Management)
JDBC	(Java Database Connectivity)
JSON	(JavaScript Object Notation)
KB	(Knowledge Base)
KCS	(Knowledge-Centered Service)
KPIs	(Key Performance Indicators)
LDAP	(Lightweight Directory Access Protocol)
MAM	(Mobile Application Management)
MDM	(Mobile Device Management)
OLA	(Operations Level Agreement)
OOAD	(Object Oriented Analysis and Design)
OOB	(Out of the Box)
PaaS	(Platforms-as-a-Service)

PL/SQL	(Procedural Language for SQL)
PPM	(Project Portfolio Management)
PPM	(Project Portfolio Management)
RC	(Risk & Compliance)
REST	(RESTful)
SaaS	(Software-as-a-Service)
SAM	(Software Asset Management)
SAML	(Security Assertion Markup Language)
SCCM	(System Center Configuration Manager)
SDLC	(Software Development Life Cycle)
SecOps	(Security Operations)
SIR	(Security Incident Response)
SLA	(Service Level Agreement)
SM	(Service Mapping)
SNPI	(ServiceNow Platform Implementation)
SOAP	(Simple object access protocol)
SP	(Service Provider)
SQL	(Structured Query Language)
SSO	(Single Sign-On)
UAT	(User Acceptance Testing)
UI	(User Interface)
URIs	(Uniform Resource Identifiers)
URL	(Uniform Resource Locators)
UX	(User Experience)

VA	(Virtual Agents)
VMs	(Virtual Machines)
VR	(Vulnerability Response)
VRM	(Vendor Risk Management)
W3C	(World Wide Web Consortium)
WebDAV	(Web Distributed Authoring and Versioning)
WSDL	(Web Service Description Language)
XML	(Extensible Markup Language)
XSLT	(Extensible Stylesheet Language Transformations)